Khmelnik S.I.

THE ENERGETICS OF TESLA TRANSFORMER

Second Edition

Israel   2012

Copyright © 2012 by Solomon I. Khmelnik

All right reserved. No portion of this book may be reproduced or transmitted in any form or by any means, electronic or mechanical, without written permission of the author.

Published by "MiC" - Mathematics in Computer Comp.
BOX 15302, Bene-Ayish, Israel, 60860
Fax: ++972-8-8691348
Email: solik@netvision.net.il
It is sent in a press on **02/13/2012**
Printed in United States of America,
Lulu Inc., ID 12580478
ISBN 978- 1-105-53653-3

Israel   2012

Abstract

The experiments with Tesla transformer are reviewed. It is shown that the existence of electric standing wave in the vicinity of the transformer follows directly from the Maxwell's equations. Further it is shown that the existence and propagation of the electric standing wave may be explained by energy exchange between this wave and the air.

Further a Tesla flat coil is considered. It is shown that directly from Maxwell equations there follows the existence of magnetic standing wave in the coil's vicinity. The existence and propagation of magnetic standing wave is also explained by energy exchange between the wave and the air.

As a consequence, it is shown that the heat energy in the transformer vicinity may be transformed by the electric standing wave into the load energy. The existing constructions for non-fuel energy generation and wireless energy transfer are analyzed. The presented theory explains the known phenomena

Contents

Introduction \ 5
1. Some Observations and Experiments \ 5
2. Electrical Field Around the Secondary Coil of Tesla Transformer \ 9
    2.1. Disconnected Conductive Strip in Magnetic Field \ 9
    2.2. Disconnected Conductive Ring in Magnetic Field \ 14
    2.3. Disconnected Solenoid in Magnetic Field \ 16
3. Electromagnetic field around the Primary Coil of Tesla transformer \ 17
    3.1. Electromagnetic field of a strip conductor \ 17
    3.2. Solenoid with flat turns \ 19
    3.3. Electromagnetic field of a flat coil \ 19
4. Electromagnetic field around a Tesla Transformer \ 22
5. Energy-dependence of a Standing Wave \ 23
6. The Conditions of Energy-dependent electric Wave \ 25
    6.1. Electrical and magnetic polarization of the dipoles of the air \ 25
        6.1.1. General remarks \ 25
        6.1.2. Electrical polarization \ 26
        6.1.3. Magnetic polarization \ 28
    6.2. Catalyzation of Heat Processes \ 30
    6.3. Temperature in area the standing electric Wave \ 33
    6.4. About the propagation speed of the area of existence of a standing wave in the air \ 34
    6.5. Transformation of the Wave's Electric Energy into Heat Energy of the Environment \ 36
    6.6. The power density of the standing wave \ 37
7. General Scheme of Energy Transformation Process \ 38
8. The Balance of Energy and Power \ 39
9. Conclusions \ 40
10. Tesla Transformer as a fuel-free Energy Generator \ 43
11. Energy Transfer by Tesla Flat Coils \ 48
References \ 50

Introduction

The Tesla transformer is widely known, as well as the numerous and poorly explained phenomena related to it. Tesla himself used the concept of ether for their explanation. The modern physics does not accept such explanations. Thus many unexplained phenomena are transferred into the field of scientific mythology (instead of stimulating scientific research). Such approach has increased nowadays, as there appear new (very attractive for practical use) inventions, using in some way or other the properties of Tesla transformer (see, for instance, [20]). This inhibits their implementation.

The author is trying to explain the mentioned phenomena, remaining within the long established physical paradigm.

1. Some Observations and Experiments

First of all we must review the known experiments with Tesla transformer.

1. In [13] we read: "Tesla had suggested that the shock wave for one moment of its explosive appearance is more like an electrostatic field than any other known electric phenomenon." The existence of electric field around Tesla transformer has been marked by many experimenters.
2. The non-radiating standing waves of unknown origin "are accumulated" round Tesla transformer, and they "crawl" about the surrounding objects [12]. The radius of their propagation may reach several kilometers (see the previous paragraph).
3. There're known experiments of field energy transmission through one wire (connected to secondary winding) [1-4] – the so called "Avramenko fork". The primary scheme of this construction (shown on Fig 1) and its technical characteristics are taken from [1]. It is reported that experimental setup included a machine generator of the power up to 100 kWt generating the voltage with frequency 8 kHz, applied to the Tesla transformer. One end of the secondary winding was free. "Avramenko fork" was connected to the second end. The Avramenko fork was a closed circuit that included two series-connected diodes D1 and D2, whose common point was connected to the wire Λ, and the load. The load in the first case was represented by condenser C and discharger P. The load in the second case was represented by several light bulbs – the resistance

R2. By this open circuit Avramenko was able to transmit an electric load of capacity 1300 Wt from generator to the load. Electrical bulbs shone brightly. The current in the wire was very small, and the thin tungsten wire (resistance R1) in the line Λ even wasn't warming.

4. During power transmission through one wire the standing waves are observed in the transmitting conductor [5]. In [3] the experiments were described which show (as the authors argue) that around a transmitting conductor there exists an electromagnetic scalar field As the field's indicator in these experiments served a metallic sphere with an iron ring which rotated during the sphere's charging – see Fig. 2. The sphere was being charged for 60 sec and discharged practically in one moment.

5. The experiments of energy transmission from Tesla transformer to fluorescent lamps are demonstrated even in school experiments and explain the radio waves propagation. But there are also known experiments in transmission through a broken wire or even without wires, or a glow from burned light bulbs. [14]

6. The winding of the secondary coil must be single-layer.

7. In Internet there are numerous videos in which a torus connected to Tesla transformer, emit long lightnings, the so-called strings. The strings are flying not from sharp endings, but directly from the windings. The most striking pictures one may see in the descriptions of Tesla's own experiments.

8. In Internet one may find information that a man located in the area of radiation feels some discomfort. The same thing was told by Tesla when he described his experiment of power transmission, There are known reports that the birds and the coastal fish were leaving the area of Tesla's experiments, of 60 km in radius.

9. There is a possibility of transmission from one Tesla transformer to an additional, near standing one. And in the additional Tesla transformer the primary winding is connected to a load, for example, to a lamp– see Fig. 3. There may be several such additional transformers, but their number does not affect the amount of consumed power of the main [15].

10. The oscillatory process in the secondary coil practically doesn't attenuate after the stopping of oscillations in the primary coil – see Fig. 4 from [15] and experiments of Kapanadze [20].

11. A measuring device not connected to anything, at a large distance from Tesla transformer (up to 1m) begins to go off-scale, not regarding the state in which its switches are. [12, 2, 14])

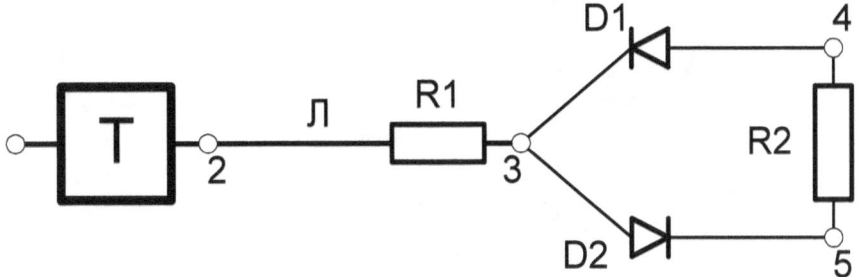

Fig. 1.

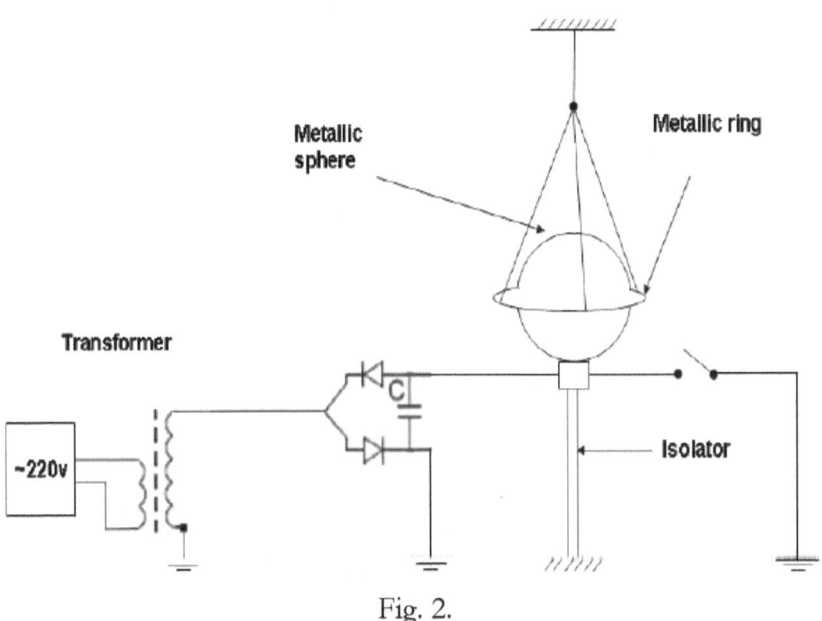

Fig. 2.

Fig. 3.

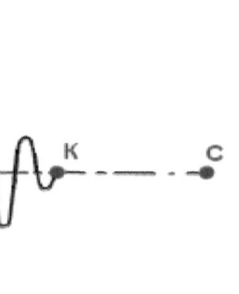

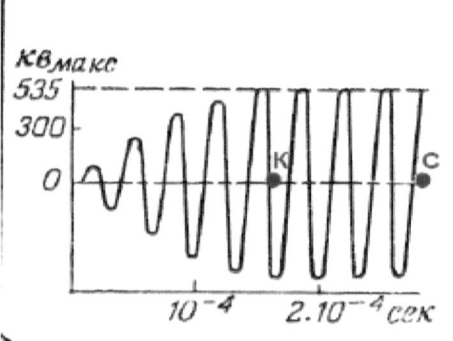

Fig. 4. Oscillograms of voltage in the Tesla coil:
a) in the primary winding b) in the secondary winding.

Various theories were suggested for explanation of these experiments and observations -see [2, 3, 4, 12]. However so far there is no universally accepted theory, and most important – not all of the observed phenomena have found explanation in these theories.

Based on the above we may assume that
- Around the transformer and the wire there emerges an electric field (without a magnetic component),
- The current in the wire is almost absent
- The speed of this field propagation is significantly less than the speed of light,
- The field presents a standing wave (it is important to emphasize that here and further we are talking about a standing wave in the space, and not about the transformer coil),
- The field possesses the energy adequate for feeding the load

The author further shows that
- The secondary winding of Tesla transformer forms a standing electric wave in the environment,
- The existence and propagation of electric standing wave is explained by the energy exchange between this wave and the air,
- The source of energy is the thermal energy of the air which is transformed by the standing electric wave into the load energy.

2. Electrical Field Around the Secondary Coil of Tesla Transformer
2.1. Disconnected Conductive Strip in Magnetic Field

Here we shall show that Tesla transformer generates in the surrounding area standing electric wave.

It is known that "e.m.f. of electromagnetic induction are induced on all parts of closed conductive circuit if these parts are crossing the (time-variable) magnetic induction lines " [11].

If the resistance closing the part $R \Rightarrow \infty$, then the current is absent. But e.m.f exists and thus the charges on the ends of the part are not equal by value or by sign. Consequently on a disconnected part there must exist a standing wave of e.m.f. and charges, changing with time. It means that on a conductive part which is crossed by magnetic (time-variable) induction lines, and closed on the resistance $R \Rightarrow \infty$:

1) Exists time-depending function of charges density distribution.
2) The current is absent.

Thus in Maxwell equations for the description of electromagnetic field generated by such a part, the currents must be absent, but changing charges must be present.

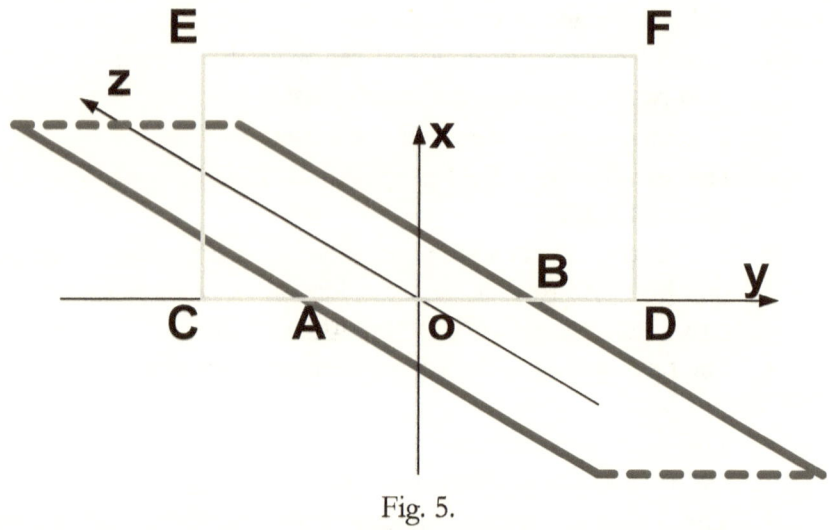

Fig. 5.

For the description of this phenomenon let us consider first a long enough conductive disconnected strip (see Fig. 5), in which a magnetic sinusoidal field (directed along the axis oy) creates E.M.F. The current in this strip is absent, but the density distribution of charges along the strip is time-variable. Thus, around such strip an electromagnetic field is generated. This field is formed by variable electric charges. Будем полагать (the grounds for this are considered further),We shall assume that the distribution function of electric charges along the strip has the following form:

$$\rho(x, y, z, t) = \rho_o \text{Chd}(\beta y) \text{Shd}(\gamma z) \lambda'(x) e^{i\omega t}, \quad (1)$$

where
- the ox axis is directed perpendicularly to the strip plane),
- the oy axis is directed across the strip,
- the oz axis is directed along the strip,
- $\text{Chd}(\beta y)$ - hyperbolic cosine function, determined along the strip width – on the interval $y \in (-R, R)$, and $2R = \overline{AB}$ - see Fig. 5;
- $\text{Shd}(\gamma z)$ - hyperbolic sine function, determined on the interval $z \in (-L, L)$, where $2L$ is the strip length;

- $\lambda'(x)$ - Dirac function,
- ρ_0, β, γ known coefficients,
- ω - angular frequency
- i - the imaginary unit.

The use of Dirac function $\lambda'(x)$ is due to the fact that at high frequencies the charges are concentrated on the surface of the conductor (skin effect).

Fig. 6 sows the functions $\text{Chd}(\beta y)$ and $\text{Shd}(\gamma z)$. The grounds for formula (1) consist in the fact that the function $\text{Chd}(\beta y)$ models uneven distribution of the charges by the strip width. The function $\text{Shd}(\gamma z)$ models uneven distribution of the charges by the strip length, which with such distribution create voltage – E.M.F. between the ends of the strip. In the further conclusions the following feature of these functions are being used :

$$\text{Chd}(w) = \frac{d(\text{Shd}(w))}{dw}, \quad \text{Shd}(w) = \frac{d(\text{Chd}(w))}{dw}, \quad (1.0)$$

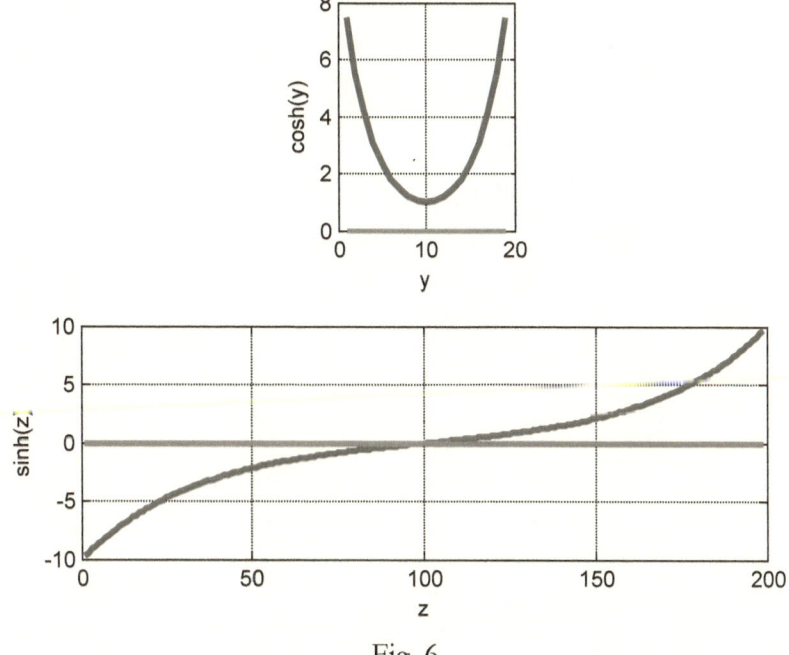

Fig. 6.

For further discussion it is important to note that ρ_0 is a function of the current I_1 in the primary coil of Tesla Transformer: the current induces e.m.f., which, in its turn, forms the charges. We can assume that

$$\rho_0 = \rho_{00} I_1, \qquad (1a)$$

where ρ_{00} – constant.

The fields in this case are monochromatic and may be presented in a complex form [7]. The Maxwell equations for monochromatic fields with regard to amplitude values for our case assume the form [7]:

$$
\begin{aligned}
&1.\ \frac{\partial H_z}{\partial y} - \frac{\partial H_y}{\partial z} - \omega\varepsilon E_x + \vartheta \frac{d\varphi}{dx} = 0 \\
&2.\ \frac{\partial H_x}{\partial z} - \frac{\partial H_z}{\partial x} - \omega\varepsilon E_y + \vartheta \frac{d\varphi}{dy} = 0 \\
&3.\ \frac{\partial H_y}{\partial x} - \frac{\partial H_x}{\partial y} - \omega\varepsilon E + \vartheta \frac{d\varphi}{dz} = 0 \\
&4.\ \frac{\partial E_z}{\partial y} - \frac{\partial E_y}{\partial z} + \omega\mu H_x = 0 \\
&5.\ \frac{\partial E_x}{\partial z} - \frac{\partial E_z}{\partial x} + \omega\mu H_y = 0 \\
&6.\ \frac{\partial E_y}{\partial x} - \frac{\partial E_x}{\partial y} + \omega\mu H_z = 0 \\
&7.\ \frac{\partial E_x}{\partial x} + \frac{\partial E_y}{\partial y} + \frac{\partial E_z}{\partial z} - \frac{\rho}{\varepsilon} = 0 \\
&8.\ \frac{\partial H_x}{\partial x} + \frac{\partial H_y}{\partial y} + \frac{\partial H_z}{\partial z} = 0
\end{aligned}
\qquad (2)
$$

Here

μ - permeability,
ε - dielectric permittivity,
φ - electric scalar potential,
ϑ - electro conductivity

These equations may be written also in this form:

$$\text{rot}(H) - \omega\varepsilon E + \theta \cdot \text{grad}(\varphi) = 0, \tag{3}$$
$$\text{rot}(E) + \omega\mu H = 0, \tag{4}$$
$$\text{div}(E) - \rho/\varepsilon = 0, \tag{5}$$
$$\text{div}(H) = 0. \tag{6}$$

Similar problems for exploring analogical equation system (up to notation and technical interpretation) were solved in [6, 8, 9]. The general method is indicated in [8]. Using it, we may find the solution of Maxwell equations for amplitudal values for given charges

$$\rho(x,y,z) = \rho_0 \text{Chd}(\beta y)\text{Shd}(\gamma z)\lambda'(x) \tag{7}$$

For $x > 0$ the solution looks as:

$$E_x(x,y,z) = e_x \text{Chd}(\beta y)\text{Chd}(\gamma z)\cos(\chi x), \tag{8}$$
$$E_y(x,y,z) = e_y \text{Chd}(\beta y)\text{Shd}(\gamma z)\sin(\chi x), \tag{9}$$
$$E_z(x,y,z) = e_z \text{Shd}(\beta y)\text{Chd}(\gamma z)\sin(\chi x), \tag{10}$$
$$\varphi(x,y,z) = \varphi_0 \text{Chd}(\beta y)\text{Shd}(\gamma z)\sin(\chi x), \tag{10a}$$

$$e_x = \frac{\rho_0}{\varepsilon}, \tag{11}$$

$$e_y = \frac{\gamma}{\chi} e_x, \tag{12}$$

$$e_z = \frac{\beta}{\chi} e_x, \tag{13}$$

$$\varphi_0 = \frac{\omega}{\chi} \rho_0, \tag{14}$$

$$\chi = \sqrt{\beta^2 + \gamma^2}. \tag{15}$$

Fig. 7 shows the example of solution for

$$\rho_0 = 10^{-7}, \quad \beta = 90, \quad \gamma = 110, \quad \omega = 1000.$$

The following functions are shown:

$$E_x(x) = e_x \cos(\chi x), \tag{16}$$
$$E_y(x) = e_y \sin(\chi x), \tag{17}$$
$$E_z(x) = e_z \sin(\chi x), \tag{18}$$

$$\varphi(x) = \varphi_o \sin(\chi x). \qquad (19)$$

It may be seen that there are oscillation of these values along the OX axis, perpendicular to the stripe plane. At the same time these value are oscillating with time (according to the conditions of the problem). It means that <u>a standing electric wave (without magnetic component)</u> appears around the stripe. The conditions of the existence of such wave will be discussed below.

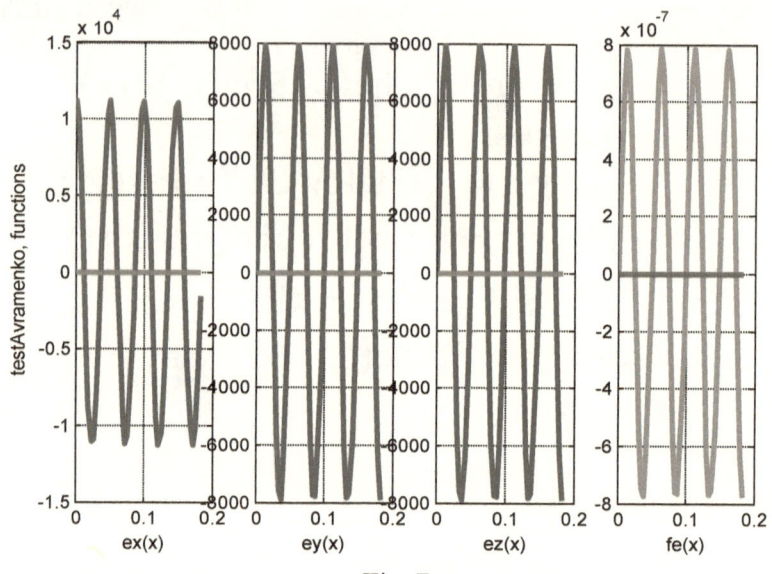

Fig. 7.

2.2. Disconnected Conductive Ring in Magnetic Field

Let us consider now a conductive disconnected strip coiled into a ring, <u>but remaining disconnected</u> – see Fig. 8. Radius of the ring will be denoted as R (interval oa on Fig. 8). Apparently, in this case instead of Cartesian coordinates x, y, z the cylindrical coordinates r, y, φ should be used. Further the electrical potential further will be denoted as φ' (unlike the previous)

Formally the transformation of Maxwell equations from Cartesian to cylindrical coordinates may be performed according the rule [8, 9]:

- the coordinates are re-denoted as:
$$x \Rightarrow r, \quad y \Rightarrow y, \quad z \Rightarrow r \cdot \varphi, \qquad (1)$$

- the coordinates are re-denoted as:
$$\frac{\partial E}{\partial x} \Rightarrow \frac{1}{r} \cdot \frac{\partial (rE)}{\partial r}, \quad \frac{\partial E}{\partial y} \Rightarrow \frac{\partial E}{\partial y}, \quad \frac{\partial E}{\partial z} \Rightarrow \frac{1}{r} \cdot \frac{\partial E}{\partial \varphi}. \qquad (2)$$

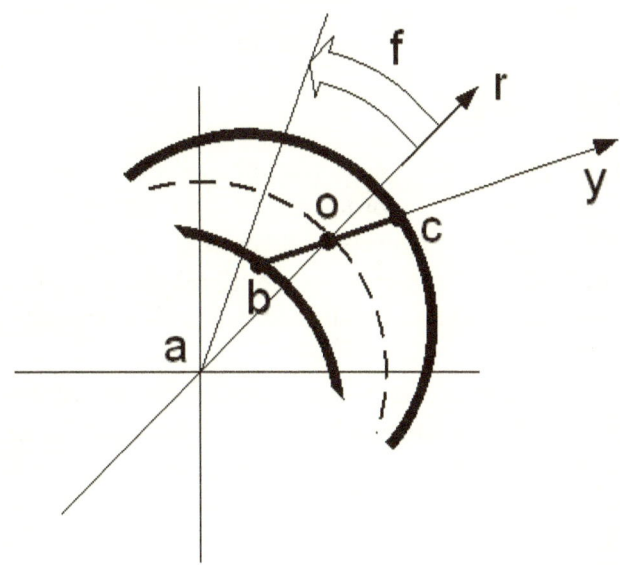

Fig. 8.

We shall assume that the density distribution function may be presented as

$$\rho(r,\varphi,y) = \rho_o \text{Chd}(\beta y)\text{Shd}(\gamma\varphi)\lambda'(R). \tag{3}$$

Then the solution of Maxwell equations will be:

$$E_r(r,\varphi,y) = \frac{e_x}{r}\text{Chd}(\beta y)\text{Ch}(\gamma\varphi)\cos(\chi(r-R)), \tag{4}$$

$$E_\varphi(r,\varphi,y) = \frac{e_z}{r}\text{Shd}(\beta y)\text{Chd}(\gamma\varphi)\sin(\chi(r-R)), \tag{5}$$

$$E_y(r,\varphi,y) = \frac{e_y}{r}\text{Chd}(\beta y)\text{Shd}(\gamma\varphi)\sin(\chi(r-R)), \tag{6}$$

$$\varphi'(r,\varphi,y) = \frac{\varphi'_o}{r}\text{Chd}(\beta y)\text{Shd}(\gamma\varphi)\sin(\chi(r-R)). \tag{7}$$

Gradient of the electric field's scalar potential along the *or* axis will have the form:

$$G_r(r,\varphi,y) = \frac{d(\varphi'(r,\varphi,y))}{dr} =$$
$$= \frac{\chi\varphi'_o}{r}\text{Chd}(\beta y)\text{Shd}(\gamma\varphi)\cos(\chi(r-R)) \tag{8}$$

Hence it follows that the solution of this problem in cylindrical coordinates differs from the solution in Cartesian coordinates by a factor

$$\xi = \frac{R}{r}, \quad r \geq R. \qquad (9)$$

This means that in Cartesian coordinates we have undamped spatial oscillations along the coordinate x, and in cylindrical coordinates –damped according to hyperbolical law oscillations along the coordinate r. The functions of electrical strength and the potential along the axis or has the sinusoidal form with monotonically decreasing amplitude.

2.3. Disconnected Solenoid in Magnetic Field

Let us apply the obtained above results for the consideration of Tesla transformer functioning. A disconnected secondary winding may be identified with a multitude of disconnected coils in a variable magnetic field. The electric field of such a construction, as follows from the previous, is <u>a standing electric wave</u>.

On the winding for $r = R$ there exists a strength (3.4), potential (3.7) and potential's gradient (3.8) or

$$E_{ro}(\varphi, y) = \frac{e_x}{R} \operatorname{Chd}(\beta y) \operatorname{Ch}(\gamma \varphi), \qquad (1)$$

$$\varphi'_o (\varphi, y) = 0, \qquad (2)$$

$$G_{ro}(\varphi, y) = \frac{\chi \varphi'_o}{R} \operatorname{Ch}(\beta y) \operatorname{Sh}(\gamma \varphi). \qquad (3)$$

Thus the secondary coil of Tesla transformer induces a standing electric wave without magnetic component. The formation is performed with current I_1 flowing in the primary coil of Tesla transformer в первичной катушке трансформатора Тесла. According to (2.1a, 2.11) it can be argued that for fixed coordinates the intensity, the potential and the gradient are known functions I_1. So the Tesla transformer as a whole is described by an equations system relating the variables

$$I_1, \ E_r(r,\varphi,y), \ E_y(r,\varphi,y), \ E_\varphi(r,\varphi,y), \ \varphi'(r,\varphi,y).$$

3. Electromagnetic field around the Primary Coil of Tesla transformer

3.1. Electromagnetic field of a strip conductor

Let us assume that the conductor in which an alternating current j with circular frequency ω, has the form of infinite strip along the coordinate z – see Fig. 5. Then

- All derivatives with respect to z become equal to zero,
- The intensity $H_z = 0$,
- The derivative of electrical scalar potential φ along a certain axis, equal to the current's projection on this axis, exists only along the axis oz.

Then the Maxwell equations system (2.1.2) takes the form

$$
\begin{array}{ll}
3. & \dfrac{\partial H_y}{\partial x} - \dfrac{\partial H_x}{\partial y} - \omega \varepsilon E_z + \vartheta \dfrac{d\varphi}{dz} = 0 \\[6pt]
4. & \dfrac{\partial E_z}{\partial y} + \omega \mu H_x = 0 \\[6pt]
5. & -\dfrac{\partial E_z}{\partial x} + \omega \mu H_y = 0 \\[6pt]
8. & \dfrac{\partial H_x}{\partial x} + \dfrac{\partial H_y}{\partial y} = 0
\end{array}
\qquad (1)
$$

This system of 4 equations from three variables is overdetermined. But the equation (1.8) follows from (1.4, 1.5) and therefore, one of the three last equations may be excluded.

If J is the projection of density of current j to the plane xoy,

$$ J = -\vartheta \frac{d\varphi}{dz}. \qquad (2) $$

Thus we shall rewrite (1.3) as follows:

$$ \frac{\partial H_y}{\partial x} - \frac{\partial H_x}{\partial y} - \omega \varepsilon E_z - J = 0. \qquad (3) $$

Let the distribution function of current density is
$$J(x, y) = J_o \text{Chd}(\beta y) \lambda'(x), \qquad (4)$$
where J_o, β are known coefficients. The function $\text{Chd}(\beta y)$ (see Fig. 6) is approximating rather closely the real distribution function of alternate current along the conductor's width (determined as skin-effect). Practically,
$$\beta \approx 1/S, \qquad (5)$$
where S is the strip width.

So, the electromagnetic field in the vicinity of stripe conductor is described by equations system (1.4, 1.5, 3, 4), where the variables are the amplitudes of complex intensities $H_x(x,y), H_y(x,y), E_z(x,y)$.

Using the method described in [8], we can show that the solution of these equations for $x > 0$ will be the following:
$$H_x(x,y) = h_x \text{Shd}(\beta y) \sin(\chi x), \qquad (6)$$
$$H_y(x,y) = h_y \text{Chd}(\beta y) \cos(\chi x), \qquad (7)$$
$$E_z(x,y) = e_z \text{Chd}(\beta y) \sin(\chi x), \qquad (8)$$
$$h_x = J_o, \qquad (9)$$
$$h_y = -h_x, \qquad (10)$$
$$e_z = -\frac{\omega \mu}{\beta} h_x, \qquad (11)$$
$$\chi = \beta. \qquad (12)$$

Example 1. Fig. 9 shows the result of numerical solution of Maxwell equations for this problem. In this example it is assumed that $\omega = 10^5$, $\beta = 500$, $J_o = 25000$. The wavelength along axis ox is $\lambda = 2\pi/\chi = 2\pi/\beta = 0.014 m$.

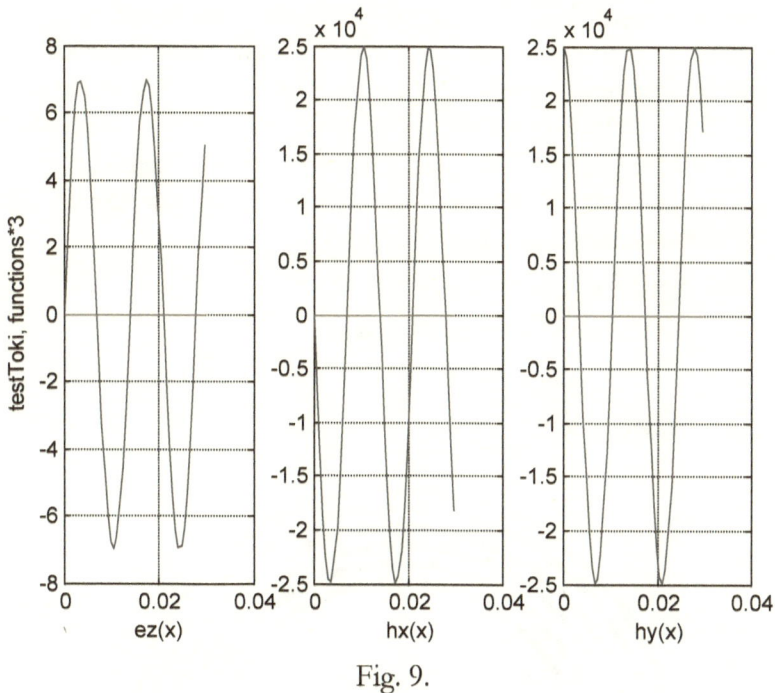

Fig. 9.

Evidently, there are fluctuations of intensities along the axis ox, perpendicular to the strip's plane. At the same time these values fluctuate with the time (by the problem's condition). It means that around the strip there appears a <u>standing electromagnetic wave</u>. It can be seen that the electrical component of this wave is small and can be ignored, In future we shall talk about a standing magnetic wave around a flat (disk) conductor. The conditions for such wave's existence will be discussed below.

3.2. Solenoid with flat turns

Let us now consider a solenoid wound with one layer of flat conductor in which alternate current flows. Reasoning exactly as before, one may note that along the radius of such solenoid there exist oscillations along coordinate r, fading according to hyperbolic law. The function of magnetic intensity along the axis or has the form of sinusoid with monotonically decreasing amplitude.

3.3. Electromagnetic field of a flat coil

Let us consider now a flat coil – the so called Tesla coil. We shall assume that it is wound by one layer of flat conductor in which an alternate current flows. To describe the electromagnetic field of such coil we must turn to cylindrical coordinate system of another kind (differing from the previous one) – in this case instead of Cartesian coordinates x, y, z we should consider cylindrical coordinates x, r, φ. Formally the transformation of Maxwell equations from Cartesian to cylindrical coordinates can be performed according to the following rule [8]:

- the coordinates are re-denoted as:
$$x \Rightarrow x, \quad y \Rightarrow r, \quad z \Rightarrow r \cdot \varphi, \tag{1}$$

- the derivatives are redefined as:
$$\frac{\partial H}{\partial x} \Rightarrow \frac{\partial H}{\partial x}, \quad \frac{\partial H}{\partial y} \Rightarrow \frac{1}{r} \cdot \frac{\partial (rH)}{\partial r}, \quad \frac{\partial H}{\partial z} \Rightarrow \frac{1}{r} \cdot \frac{\partial H}{\partial \varphi}. \tag{2}$$

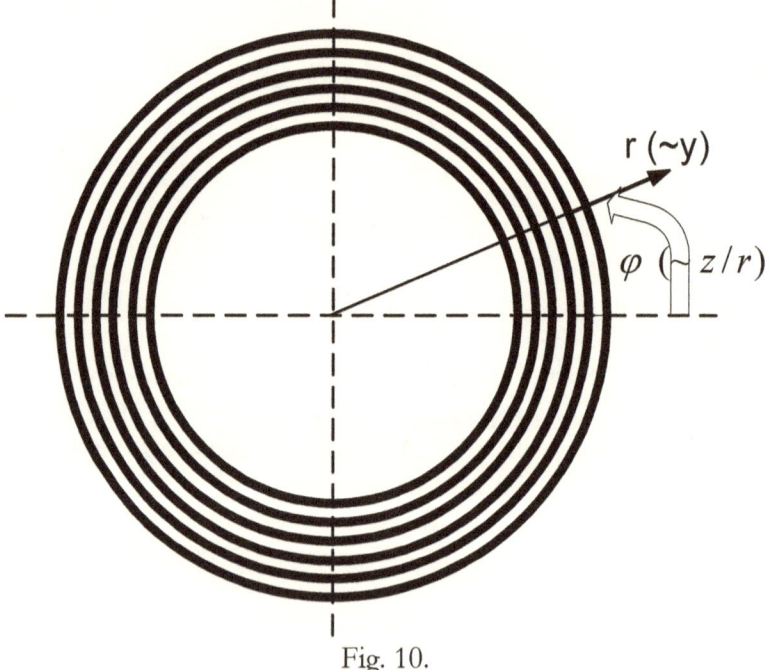

Fig. 10.

This transformation is illustrated on Fig. 10, where the axis ox is perpendicular to the coil's plane, the axis $oy \Rightarrow or$ us directed along the circle's radius, axis $oz \Rightarrow r \cdot \varphi$ - arc of the coil.

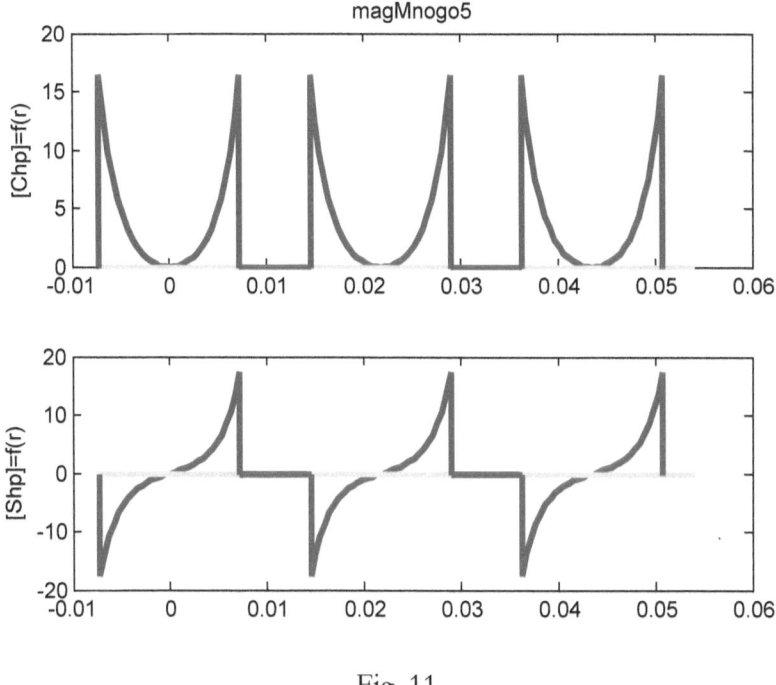

Fig. 11.

In our case we may also present the distribution function of current density in the form (3.1.4), namely:
$$J(x,r) = J_o \text{Chp}(\beta r) \lambda'(x), \qquad (3)$$
But the function $\text{Chd}(\beta y)$, depicted on Fig. 6, in our case takes the form of function $\text{Chp}(\beta r)$ - see Fig. 11, where the functions $\text{Chp}(\beta r)$ and $\text{Shp}(\beta r)$ are shown for a three times winded coil These functions also possess the quality (2.1.1.0).

The solution of Maxwell equations in this case is similar to the case of one strip in the Section 3.1, only that the functions $\text{Chd}(\beta y)$ and $\text{Shp}(\beta y)$ should be changed to functions $\text{Chp}(\beta r)$ and $\text{Shp}(\beta r)$.

So, in this case we have undamped spatial oscillations <u>along the coordinate</u> x, which is perpendicular to the coil's plane. Thus a standing magnetic wave appears above the flat coil.

Let us note also that the same solution is obtained for a flat coil with bifilar winding – the changes are only in the functions $\text{Chp}(\beta r)$ and $\text{Shp}(\beta r)$ - see Fig. 12. These functions also have the feature (2.1.1.0).

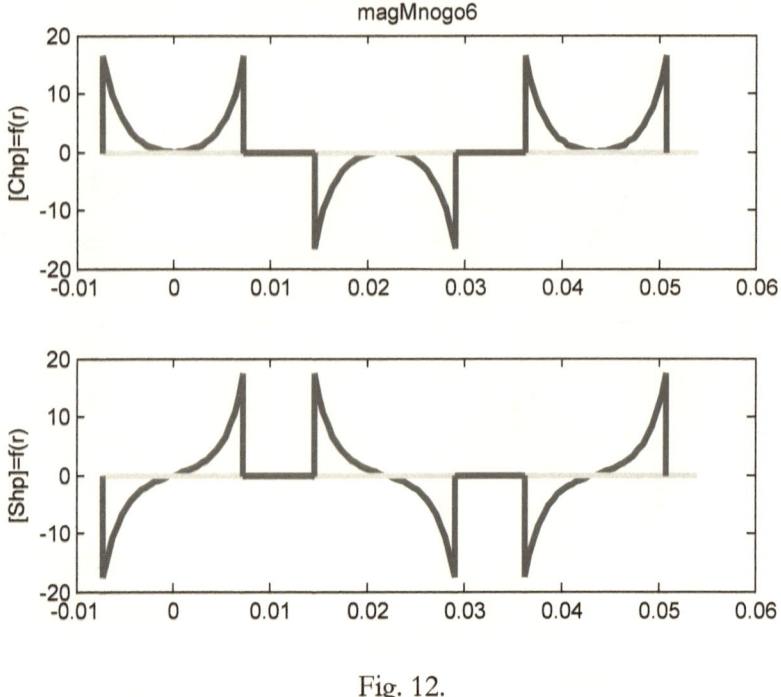

Fig. 12.

4. Electromagnetic field around a Tesla Transformer

Let us consider a Tesla transformer with flat primary coil – see, for instance, Fig. 13 from [33]. The arrows show the vectors of intensity of the standing wave created by the coils of Tesla transformer – the intensity of electrical field **E** of the secondary coil and the intensity of the magnetic field **H** of the primary coil. These fields are created by different sources and hence they are independent. It means that these fields do not exchange energies. They exist, as it will be shown further, due to energy exchange with the environment. Both of these fields are influencing the dipole in coordination. Fig. 13 shows the position of Dipole **D**, polarized by the fields **E** and **H**.

Fig. 13.

5. Energy-dependence of a Standing Wave

Undoubtedly, a question arises – how can an electric or magnetic wave exist without (respectively) a magnetic or electrical component (because in known cases an electromagnetic wave lasts due to energy exchange between the electric and magnetic components)?

In [10] is shown experimentally that there exist electric waves (without a magnetic component). In the area of such wave the decrease of temperature up to 7 grades is observed. In [9] it is proved that such wave may exist only in the conditions of energy exchange with the environment. It is shown that in such wave <u>a magnetic polarization of the air dipoles</u> is observed, consisting in the fact that the dipoles are polarized by Lorenz forces in the direction of the vector of speed with which they enter into the area of this wave. Further it is shown that such polarization substantially limits the degree of freedom of the air molecules, and this leads to decrease of internal energy of the air, and, as a consequence, to decrease in its temperature. The changing wave's energy summing with the change in air's internal energy satisfy the energy conservation law. The conditions of satisfying the energy conservation law are also the conditions of this wave's existence. It is shown that the decrease in temperature around the wave is a result of this condition. This phenomenon (as was noted above) is observed in experiments [10].

There exists a certain speed of propagation of this wave in the air.

Similarly we can show that the electric standing wave can exist only in the conditions of energy exchange with the environment. In such wave an <u>electric polarization of dipoles</u> must be observed. Such polarization substantially limits the degree of freedom of the air molecules, and this leads to decrease of internal energy of the air, and, as a consequence, to decrease in its temperature. The changing energy of the wave summing with the change in air's internal energy satisfy the energy conservation law. The decrease in temperature around the wave should be a result of this condition. There exists a certain speed of propagation of this wave in the air, which is shown in the described above experiments. By analogy with this experiment it may be expected that the area of the wave existence depends on the load intensity.

So, the standing electrical or/end magnetic field is created on account of the transformer's energy, it exists on account of energy exchange with the environment, it propagates and is able to transmit a part of its energy to the load.

6. The Conditions of Energy-dependent Wave
6.1. Electrical and magnetic polarization of the dipoles of the air
6.1.1. General remarks

Let us consider Figure 14, where the notations are as follows:

L - length of dipole,

q - charge of dipole,

$\overline{L}$ - vector of dipole directed from "−" to "+",

$\overline{E}$ - intensity of electric field,

$\overline{H}$ - intensity of magnetic field directed perpendicularly to the Figure's plane,

$\overline{V}$ - speed of dipole center movement,

$\overline{F}$ - Lorentz's force affecting on a unit charge moving in a magnetic field,

α - angle between the dipole vector $\overline{L}$ and the intensity $\overline{E}$.

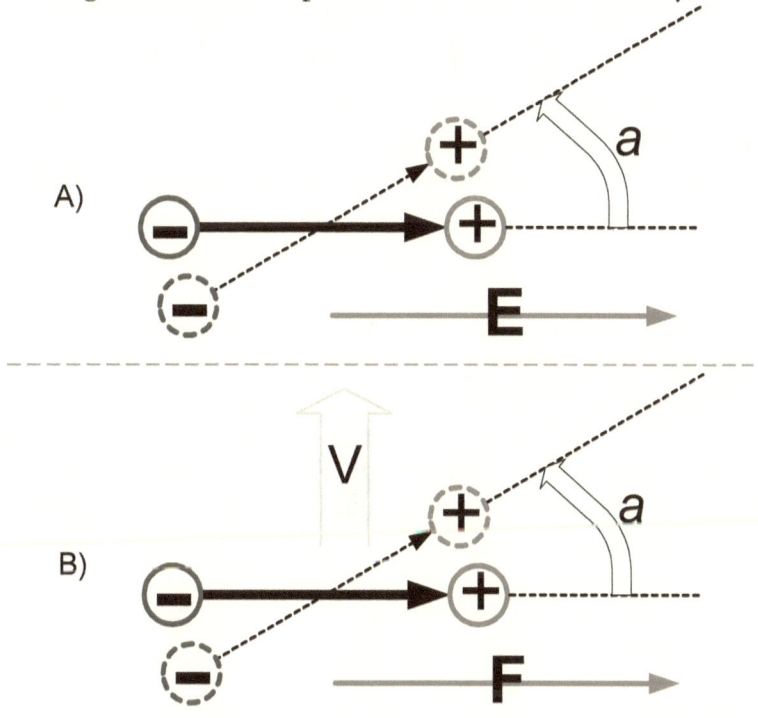

Fig. 14. A) Electric dipole in an electric field,
B) Electric dipole moving with the V in a magnetic field

Fig. 14 shows the process of electric dipole polarization in the standing magnetic and electric waves that are formed in the Tesla transformer – see also Fig. 13. It is assumed that the vector of intensity **H** of the magnetic field is directed perpendicularly to the figure's plane, and the vector of intensity of the electric field **E** is directed as it is shown on the figure.

6.1.2. Electrical polarization

The dipole's electric moment is

$$\overline{p} = q\overline{L}. \qquad (1)$$

The dipole's stable state corresponding to the minimum of its potential energy, is the position $\alpha = 0$, for which the dipole's vector $\overline{L}$ and intensity $\overline{E}$ are parallel and the torque is equal to 0 – see Fig. 1. The dipole orientation along the vector of intensity $\overline{E}$ is called the <u>electrical polarization</u> [16]. The work performed by the field in the process of dipole's moving into the state of stable equilibrium, is equal to [16]

$$W_e = (\overline{p}, \overline{E}), \ W_e = qLE\cos(\alpha). \qquad (2)$$

This is the variation of dipole's potential energy in the process of its orientation. The work performed by the field in the process of dipole orientation is performed on account of the variation of internal energy of a dielectric W_T. When the angle α changes, the potential energy of dipole W_e also changes, as well as the internal energy of the dipole W_T (foe the air if it is regarded as an ideal gas, this is the energy of the molecules heat motion), but their sum according to the energy conservation law remains constant - see Figure 9.

Thus, the <u>electric polarization of dipoles reduces the internal energy of dielectric</u> but the number of oriented dipoles increases with the increase of intensity. Hence, <u>the internal energy of dielectric decreases with the intensity growth.</u>

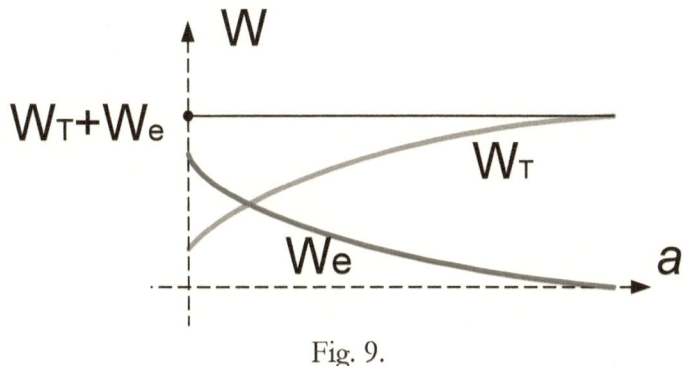

Fig. 9.

The polarization energy density, as it is known from [16], is equal to

$$W_e = \varepsilon_o \chi_e E^2 / 2, \qquad (3)$$

where

χ_e is the orientation polarizability by electric field,

ε_o - electrical permittivity of vacuum

The electrical polarization is expressed in different ways [16]:

$$P_e = p n_e = \varepsilon_o \chi_e E, \qquad (4)$$

where n_e is the number of polarized dipoles per cubic meter of air. Hence we may find:

$$n_e = \varepsilon_o \chi_e E / p, \qquad (5)$$

$$W_e = p n_e E / 2. \qquad (6)$$

Thus, <u>the energy of dipoles eectrical polarization is determined by the formula (6).</u>

We shall assume that the relative number of polarized dipoles in a cubic meter of air is

$$\overline{n_e} = n_e / n_o, \qquad (7)$$

where n_o is the number of dipoles in a cubic meter of air in proportion with electric intensity, i.e.

$$\overline{n_e} = \xi_e E. \qquad (8)$$

The value of coefficient ξ_e will be evaluated further.

The relative permittivity is

$$\varepsilon = 1 + \chi_e \qquad (9)$$

6.1.3. Magnetic polarization

Let us consider now the process of electric dipole polarization in a magnetic field [9]. This process is depicted on the Figure 1b. If the dipole's center is moving with the speed $\overline{V}$ (in this case it is the speed of heat motion), then the Lorentz force will be affecting each charge of the dipole. Rated at a unit positive charge this force will be equal to

$$\overline{F} = \mu_o \overline{H} \times \overline{V}, \tag{8}$$

where $\overline{H}$ is the magnetic field intensity, μ_o -the magnetic constant. The forces affecting each charge of the dipole will create a rotating momentum and turn the dipole. When the dipole will turn so that the dipole vector $\overline{L}$ will be perpendicular to the speed vector $\overline{V}$, then the Lorentz forces will be directed in opposite directions and their rotating momentum will be equal to zero. Dipole will assume the state of stable equilibrium (the dipole's movement in the process of its polarization will be considered in more detail in [9]). The above mentioned force $\overline{F}$ will be similar in its effect to the effect of intensity $\overline{E}$ on the polarization in an electric field. Based on this analogy the variation of potential energy of an electric dipole moving in magnetic field is expressed by a formula similar to the formula (2)

$$W_h = (\overline{p}, \overline{F}), \quad W_h = qLF \cos(\alpha). \tag{9}$$

The angle α is counted from the state of stable equilibrium for which $\alpha = 0$.

Thus, the orientation of dipole perpendicularly to the speed vector $\overline{V}$ will be called <u>the magnetic polarization of electric dipole</u> (by analogy with electric polarization. By analogy with the above said we may declare that <u>for a certain speed of dipoles movement with the increase of magnetic intensity the internal energy of dielectric decreases</u>.

We must note that contrary to the case of dipoles electric polarization, here the dipoles are oriented in different directions due to chaotic molecules motion), and so there does not arise a summary magnetic moment similar to the vector $\overline{P}_e$). Because of this the described magnetic polarization can not be observed experimentally. We must also note that there also takes place the magnetic polarization of the molecules of the air as a paramagnetic, parallel to vector $\overline{H}$. But this effect is quite insignificant, and so we do not consider it here.

The density of magnetic polarization energy will be expressed by analogy with formula (6)

$$W_h = pn_h F, \qquad (10)$$

where n_h is the number of polarized dipoles in a cubic meter of air.

We shall assume that the relative number of magnetically polarized dipoles in a cubic meter of air is

$$\overline{n_h} = n_h / n_o, \qquad (11)$$

where n_o is the number of dipoles in a cubic meter of air in proportion with magnetic intensity, i.e.

$$\overline{n_h} = \xi_h H. \qquad (12)$$

The value of coefficient ξ_h will be evaluated further (its dimension is m\A).

Combining (8) and (10), we find:

$$W_h = pn_h(\mu_o HV). \qquad (13)$$

So we see that the magnetic polarization energy of dipoles is determined by formula (13).

The thermal processes in a magnetic standing wall are discussed in detail in [9].

6.2. Catalyzation of Heat Processes

As the considered wave really exists, it is necessary (for the law of energy conservation fulfillment) to discover the additional energy source. Due to the energy conservation law the sum of energy density of the electric field and the energy of dipoles polarization must be equal to zero, i.e.

$$\frac{\varepsilon_o \varepsilon \cdot E^2}{2} + W_e = 0 \qquad (1)$$

or

$$\frac{\varepsilon_o(1+\chi_e)E^2}{2} + \varepsilon_o \chi_e E^2 / 2 = 0. \qquad (2)$$

This equality DOES NOT hold. Then there must exist some other source of energy. Further we shall show that the thermal energy of the air serves as such source. This energy in some way transforms into the energy of polarization of the air molecules, and, as it seems, we still have to use the formula (2). The explanation of such contradiction may be found only in the fact that the orientational polarizability in the conditions of time-variable standing wave differs from the polarizability determined by permanent electric field. The condition (6.1.9), used in (2), is also violated. Further we assume that the permittivity of the air always is equal to ε_0.

To avoid separate discussion for the energy change in the molecules of oxygen and nitrogen, we shall consider an imaginary molecule of the air with averaged parameters. The mass of air mole with such averaging is equal to $29 \cdot 10^{-3} kg$. Such simplification is frequently used in the analysis of thermal qualities of the air.

Usually in the gas various kinds of molecules movement is possible (translational movement along the three axes and rotary movement around these axes), which corresponds in general case to 6 degrees of freedom, and for the air as two-atomic gas – 5 degrees of freedom (as the rotation around the dipole axis may be ignored). In molecular physics [17] the calculations of molecules heat movement energy are performed under the assumption of energy uniform distribution by degrees of freedom.

For electrical polarization of dipoles, when dipole is kept oriented along line of force by the electric field forces, the two rotary movements are impeded. Hence the energy for two degrees of freedom is absent.

Mathematically all the said may be described as follows: when calculating the internal energy of air for dipoles electric polarization the number of degrees of freedom should be reduced by 2, and for magnetic polarization – by 2.

Thus, electrically polarized dipole loses two degrees of freedom. If before polarization the heat energy of dipole was equal to $5kT/2$, then after electric polarization it will be reduced by kT (where k is Bolzman's constant, T – absolute temperature of the air). So, the polarization of dipoles catalyzes the decrease of the air's internal energy.

Physically the decrease of internal energy in the process of dipoles polarization may be explained as follows: without polarization at molecules collision the law of kinetic energy conservation is being fulfilled for a system of two molecules. For collision of two polarized molecules (due to the change of speeds directions) the dipoles deviate from their state of stable equilibrium. For this some quantity of energy is needed. Consequently, the kinetic energy of molecules after the collision decreases, i.e. the internal energy of the air is reduced. Mathematically it is taken into account by the reduction of degrees of freedom.

Let us write the value of air's internal energy density reduction in the process of polarization:
$$W_T = kTn_e. \tag{3}$$
Then we shall get from (6.1.7, 3):
$$W_T = kTn_o \overline{n_e}. \tag{4}$$
Let us note that full internal energy of the air is
$$W_{TO} = 2.5kTn_o \tag{5}$$
and after <u>full</u> polarization, when $\overline{n_e} = 1$, it is reduced by
$$W_{T\max} = kTn_o. \tag{6}$$
i.e. reduced by 40%. Let us rewrite (4) in the form
$$W_T \approx \sigma \overline{n_e} \tag{7}$$
where
$$\sigma = kTn_o. \tag{8}$$
After substitution of (6.1.8) into (7) we get
$$W_T = \sigma \xi_e H. \tag{9}$$

So we see that the decrease of the air's internal energy depends on the number of oriented molecules, which, in its turn, depends of the electric intensity. It is this energy that participates in the process of energy exchange in the considered wave. The balance of energy in this case is described by the equation

$$\frac{\varepsilon_o E^2}{2} - W_T \approx 0. \tag{10}$$

or, taking into account (4),

$$\frac{\varepsilon_o E^2}{2} - kTn_o \overline{n_e} \approx 0. \tag{11}$$

or, taking into account (9),

$$\frac{\varepsilon_o E^2}{2} - \sigma \xi_e E \approx 0. \tag{12}$$

Hence, the considered wave exists if the condition (12) is fulfilled. From it we find:
$$\xi_e \approx 0.5 \varepsilon_o E / \sigma. \tag{13}$$
or

$$\xi_e \approx 2\pi\sigma_1 E. \qquad (14)$$

where
$$\sigma_1 \approx \varepsilon_o / (4\pi\sigma). \qquad (15)$$

Taking into account (6.1.7, 6.1.8) we get:
$$\overline{n_e} \approx 2\pi \cdot \sigma_1 E^2, \qquad (16)$$
$$\overline{n_h} \approx 2\pi \cdot \sigma_1 E^2 n_o. \qquad (17)$$

Example 1. For $n_o = 3 \cdot 10^{25}$, $k = 1.38 \cdot 10^{-23}$, $T = 300$, $\varepsilon_o = 8.85 \cdot 10^{-12}$, $p_e = 2 \cdot 10^{-29}$ according to (6.1.7, 6.1.8, 8, 14-17), we find: $\sigma = kTn_o \approx 10^5$, $\sigma_1 \approx \varepsilon_o/(4\pi\sigma) \approx 10^{-16}$, $\xi_e \approx 2\pi\sigma_1 E = 4 \cdot 10^{-11}$, $\overline{n_e} = \xi_e E = 4 \cdot 10^{-11} E$, $n_e = \overline{n_e} \cdot n_o \approx 10^{15} E$. If we are taking into account the reference value of orientational polarizability being $\chi_e = 5 \cdot 10^{-17}$, then we shall find $n'_e = \chi_e/p_e = 2.5 \cdot 10^{12}$. Then $\lambda = n_e/n'_e \approx 400E$. Hence it follows that the time-variable standing wave _increases the polarization of air_ $\lambda \approx 400E$ _times as compared with a permanent electric field._

6.3. Temperature in area the standing Wave

It is known [18], that the dependence of gas internal energy and its temperature T for volume unit has the following form:
$$u = sDT, \quad D = \frac{\rho R}{2M}, \qquad (1)$$

where

s – the degrees of freedom number for air molecules (for the air – two-atomic gas $s = 5$),

$\rho \approx 1.2 \text{kg/m}^3$ - the air density,

M - the gas mole mass (for the air $M = 29 \cdot 10^{-3}$ kg/mol),

$R = 8.31 \text{J}/(\text{mol} \cdot \text{K})$ - universal gas constant

Thus, the air constant is $D \approx 175 \text{J}/(\text{m}^3\text{K})$

The change of internal energy at temperature change by ΔT is determined by the formula

$$\Delta u = -sD \cdot \Delta T. \quad (2)$$

First we shall consider the temperature in the vicinity of electric standing wave. Assuming that this value is equal to the air internal energy decrease at the time of molecules polarization, and taking into account the fact, that after polarization $s = 1$, from (6.2.10) we get:

$$\frac{\varepsilon_o E^2}{2} + D \cdot \Delta T \approx 0. \quad (3)$$

From this we find

$$\Delta T = -\frac{\varepsilon_o E^2}{2D}. \quad (4)$$

or

$$\Delta T = -2.5 \cdot 10^{-14} \cdot E^2. \quad (5)$$

The intensity causing decrease of temperature of ΔT,

$$E = 6 \cdot 10^6 \sqrt{\Delta T}. \quad (6)$$

Example 2. Let $\Delta T \approx -0.0001^0$. Then the intensity $E \approx 6 \cdot 10^4$. So it follows that the temperature decrease is practically imperceptible.

Now we shall consider the temperature in the vicinity of magnetic standing wave. In this case instead of formula (3) a formula of the form [9] will be used:

$$\frac{\mu_o H^2}{2} + D \cdot \Delta T \approx 0. \quad (3)$$

From this we find

$$\Delta T = -\frac{\mu_o H^2}{2D}. \quad (4)$$

or

$$\Delta T = -\frac{B^2}{2D\mu_o}, \quad (4a)$$

$$\Delta T = -2300 \cdot B^2. \tag{4b}$$

Example 3. Let $B \approx 3 \cdot 10^{-5} T$ - This value is used below in Section 11. The corresponding intensity is $H = B/\mu_0 \approx 25\, A/m$. Using formula (4b), we shall find the temperature lowering $\Delta T \approx -0.07^0$.

6.4. About the propagation speed of the area of existence of a standing wave in the air

In the equation (6.2.10) E is the intensity amplitude, and $\overline{n_e}$ – is the average value of the relative amount of polarized dipoles in the air. Evidently, the amount of polarized dipoles in the air, being dependent on H (see (6.1.8)), varies synchronously with H. So we shall consider the amplitudinal value of relative amount of polarized dipoles and write once more (6.2.10):

$$\frac{\varepsilon_0 E^2}{2} - kTn_0\overline{n_e} \approx 0. \tag{1}$$

Let us compare (1) with the equation of balance of magnetic and electric energies of traveling wave in the vacuum

$$\frac{\mu_0 H^2}{2} - \frac{\varepsilon_0 E^2}{2} \approx 0, \tag{2}$$

which propagates with the speed of light

$$c = 1/\sqrt{\varepsilon_0 \mu_0}. \tag{3}$$

Let us rewrite (1) in the form:

$$\frac{\varepsilon_0 E^2}{2} - \frac{2kTn_0\left(\sqrt{\overline{n_e}}\right)^2}{2} \approx 0. \tag{3a}$$

Comparing (3a) with (2), by analogy we find the propagation speed of dipoles polarization wave:

$$d = 1/\sqrt{2kTn_0\varepsilon_0}. \tag{4}$$

Perhaps the , this is apparently the propagation speed of the considered longitudinal wave in the air. So, <u>the propagation speed (the extension of existence domain) of electromagnetic energy-dependent electrical standing wave in the air</u> is determined by (4).

Example 4. In normal conditions

$$n_0 = 3 \cdot 10^{25} \text{м}^{-3}, \ T = 300\text{K}, \ k = 1.38 \cdot 10^{-23}, \ \varepsilon_0 = 8.85 \cdot 10^{-12}.$$

Then from (4) we find:
$$d \approx 700 \text{ м\сек}. \tag{5}$$

Example 5. In [9] it is shown that similarly calculated <u>speed of the extension of magnetic standing wave existence area in the air</u> has the value
$$d \approx 2.5\text{м\сек}. \tag{5a}$$

In vacuum $n_0 = 0$ the existence domain of the wave "collapses" into a point. Thus, this wave can exist only in the air environment. However, here the speed of heat flow propagation which must accompany the wave propagation is not taken into account.

6.5. Transformation of the Wave's Energy into Heat Energy of the Environment

Aside from electrical polarization there exists also an adverse process. The inflow of heat energy from the environment (medium 2) caused by temperature decrease of the area existence of a wave (medium 1), depolarizes the air molecules. Depolarization means the rotation of polarized air molecules under the influence of heat movement of the surrounding molecules (striking the polarized molecule). Such rotation of a molecule – electric dipole, - creates a magnetic field. Then the energy of medium 1 grows up (at the expense of heat flow from medium 2), and the energy of the wave grows down.

Thus, there takes place an oscillatory process caused by variations of electric field intensity, and, accordingly, variations of the field instantaneous energy:
- if intensity and field instantaneous energy <u>increase</u>, then
- intensity polarizes molecules
- and reduces the instantaneous energy of medium 1,
- if intensity and field instantaneous energy <u>decrease</u>, then,
- instantaneous energy of medium 1 increases (at the expense of heat flow from medium 2), i.e. the heat movement in medium 1 is activated,
- it depolarizes the molecules

- and reduces the intensity and field instantaneous energy
- and so on

Varying energy of the wave in sum with varying internal energy of the air in medium 1 satisfy the law of energy conservation. The conditions of this law compliance are also conditions of the wave's existence.

In [9] it is shown that there exists also a similar process of transformation of the wave's magnetic energy into thermal energy of the environment.

6.6. The power density of the standing wave

It was shown above that the change of internal energy of medium 1 with temperature change by ΔT is determined by (6.3.2) for $s = 1$, i.e.

$$W_T = -D \cdot \Delta T. \tag{1}$$

For example, if in the field of electrical wave $\Delta T \approx -0.0001^0$, then $W_T \approx 0.02 \text{J/m}^3$. The energy $|W_T|$ transforms into the energy of an additional field, generated by depolarization. So, as a result of heat inflow from external medium 2 there occurs the energy increase of medium 1 and energy decrease in the electric field. On a whole this is equivalent to the transformation of the wave's energy into heat energy.

The value (1) unsigned is the density of the electric field's power, and it related to the wave's time period. Power density of magnetic field is [9]

$$P \approx -\omega W_T = \omega \cdot D \cdot \Delta T. \tag{1a}$$

For example, if in the field of electrical wave $\Delta T \approx -0.0001^0$ and $\omega = 10^5$, then $W_T \approx 0.02 \text{J/m}^3$ и $P \approx 2000 \text{Wt/m}^3$.

7. General Scheme of Energy Transformation Process

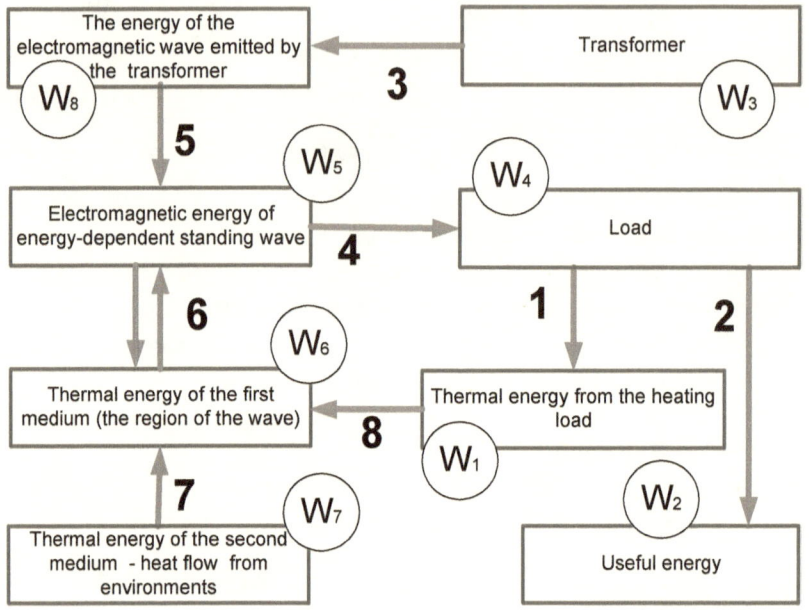

Fig. 16

[9] contains the discussion of energy processes in such a system, where exists a magnetic standing wave. Here we shall discuss the energy processes in such a system, where exists an electrical standing wave, namely, in the system with Tesla transformer.

Further we shall consider the energy processes in this system. Fig. 16 presents the scheme of energy transformation in the system. Energy of a certain type is further denoted as $\overline{W_k}$, and the volumetric density of the energy - as W_k. Energy transformation is characterized by the power transmitted from one part of the system to another. Let us denote the full power of energy $\overline{W_k}$ transformation as $\overline{P_k}$. Specific (by volume or by mass) powers denote as P_k. The arrows on Fig. 10 show the direction of power flows. In Table 1 all components are listed and the formulas for powers computation are given. Their derivation is given below.

Table 1.

κ	Part system	Energy W_k of part system	Power P_k of part system
1	Thermal energy from the heating load	W_1	Thermal energy from the heating load
2	Useful energy	W_2	Useful power
3	Transformer	W_3	Power network
4	Heater	W_4	Power of heater
5	Standing wave	W_5	
6	Medium 1	W_6	
7	Medium 2	W_7	Power of heat flow from environments
8	Emitted by the transformer (secondary coil)	W_8	Power of secondary coil

8. The Balance of Energy and Power

Looking at Figure 2, we note that in accordance with the law of energy conservation the following relations are valid. Environment 1 and wave energy exchange in the oscillatory process. Therefore, averages for the period the energy of the environment 1 and wave energy are equal, ie

$$W_6 = W_5. \qquad (1)$$

The power of thermal flow changes the energy of medium 1 and the energy of wave, i.e.

$$\frac{d(\overline{W_5})}{dt} = \overline{P_7}. \qquad (2)$$

There exist the following evident dependencies:

$$\overline{P_3} = \overline{P_8}. \qquad (3)$$

$$\overline{P_5} = \overline{P_8} - \overline{P_4} + \overline{P_7} + \overline{P_1}. \qquad (4)$$

The consumer uses a part of energy for his benefit (energy of motors, used for lighting, and so on) – we shall call this part <u>useful energy</u>. Another part of the energy received by consumer is spent useless for him

– is radiated in the form of thermal energy, for instance, the energy of lighting devices. Thus

$$\overline{P_4} = \overline{P_1} + \overline{P_2}. \tag{5}$$

From (4, 5) we find that

$$\overline{P_5} = \overline{P_3} + \overline{P_7} - \overline{P_2}. \tag{6}$$

If the energy of an electromagnetic wave remains permanent, then

$$\overline{P_2} = \overline{P_3} + \overline{P_7}. \tag{7}$$

Thus, the consumer uses the energy of transformer and environment. The power of transformer may be significantly less than the power of thermal flow. In this case it serves as a catalyst of thermal flow, whose power is used by the consumer.

If the consumer is a light bulb, then the most part of its power $\overline{P_1}$ is expended on heating and the remaining part $\overline{P_2}$ – on lighting. And $\overline{P_2}$ is only about 5% from its full power in incandescent lamps and about 10-15% - in fluorescent lamps. It means that using TT allows to reduce energy consumption by снизить 95% - 85% even in the case when the thermal flow is absent. This explains the spectacular experiments of Kapanadze [20].

9. Conclusions

The existence and propagation of electric standing wave (to be more precise, the area of its existence), energy exchange between this wave and the air allow to explain many of phenomena described at the beginning. Further there will be used their numeration in the Sector 1

1. Exactly this assumption of Tesla the author.
2. It was explained above.
3. The wire connected to the secondary winding in point 2 , keeps potential (4.2) on all its entire length, and including the opposite end – in point 3. In the vicinity of point 3 on some distance – in points 4 and 5 the potentials may differ from (4.2). Depending of the potential's polarity in point 3 the potential in point 4 (or 5) is equal to potential in point 3 (which is determined by the direction of the diode inclusion) In the other point 5 (or 4) the potential is not equal to potential in point 3. This difference of potentials is the voltage on the load of Avramenko fork. From this it follows that the wire can be soldered to any point of the secondary winding

and even simply be close to it.
4. It was explained above.
5. These phenomena are easily explained by constant potentials difference between any two points in the electric field of a standing wave.
6. This fact has been used in the mathematical proof.
7. This fact and also the plenty of strings is easily explained: the longitudinal electric field creates canals in the air, where the dipoles of the air are oriented in one direction, and the sparks are slipping.
8. Evidently, the electric sanding wave somehow affects the biological organism containing dipole molecules.
9. This is explained by the fact that the energy of electric standing wave is being used, and is being replenished by the energy of the air. First of all let us note that such devices should be extremely unstable due to the divergence of resonance frequencies of the primary and secondary Tesla transformers – for details see Section 10.
10. It was explained above.
11. The difference of potentials is present directly on the clamps of measuring element. The other elements of the device, serving only for coordination between the input voltage of the device and the input voltage of the measuring device, do not influence the above named difference of potentials.

So, the known effects follow from the presented theory as well as the effects which are to be expected, and which may confirm this theory (measurement of wave, the temperature lowering around the wave, the increase of wave volume as the load increases).

Instantaneous value of intensity for the considered wave changes in sinusoidal mode with the time. The wave's instantaneous energy (proportional to square of intensity) changes periodically from zero to a certain maximum.

When the wave's instantaneous energy grows, the instantaneous heat energy of medium 1 declines due to polarization of the air molecules. In this way occurs the transformation of heat energy into magnetic energy.

When the wave's instantaneous energy declines, the instantaneous heat energy grows. The increase of heat energy of medium 1 occurs due to the heat inflow from the external medium 2, which depolarizes the air molecules. In this way occurs the transformation of electric energy into heat energy. It is possible because there exists a difference of temperatures between medium 1 and medium 2.

Average density of a medium's energy is proportional to its temperature. Average energy density of medium 1 is less than average energy density of medium 2 by a value proportional to the difference of temperatures. This value is also equal to the average energy density of electric wave.

Full electric energy of a wave (in all domain of its existence) is equal to the energy of heat flow – just like in an ordinary electromagnetic wave magnetic energy is equal to electric energy.

The energy of heat flow from medium 2 cannot exceed the electric energy of the wave. This exceeding energy may be spent on
- the extension of the wave's existence domain (with the speed determined above);
- refilling the wave energy, if it partially transforms into other kinds of energy, for instance, into electric energy of a coil entered into the wave domain.

In the latter case the wave behaves as a heat pump. An important distinction, however, lies in the fact that for such heat pump functioning there is no need for an additional energy source.

The changing electric wave energy in the sum with changing internal energy of the air fulfill the law of energy conservation. The conditions of this law fulfillment are exactly the conditions of this wave existence. The consequence of this condition is the temperature lowering in the wave area.

The energy-dependent electromagnetic wave keeps existing as this wave exchanges energy with the environment in which this wave exists. Wave velocity for this case is defined above.

The thermal flow power $\overline{Q}$ grows till the moment when this power becomes equal to the power conveyed by the wave to the users (this power is lost in the process of area's expansion, due to the inevitable absorption of the wave's energy by the medium). With the growth of $\overline{Q}$ the radius R_{max} and the volume V of the wave area also grows. The temperature in a certain point of wave area and the induction amplitude associated with this temperature decrease with the distance from this point to the transformer coil with radius R_o [9].

Thus, the wave propagates in the direction of the intensity vector, and the value of this vector (as the amplitude of the fluctuating intensity) decreases. Such process characterizes a longitudinal wave. This is why <u>the energy depending wave is a longitudinal one</u>. Simultaneously it remains a standing wave, because the wave's nodes do not move (only their quantity increases). The wave velocity in this case is determined by the inertia of the generator (more exactly, by the speed of generator's load power change).

Note that the theory of electromagnetic waves admits the existence of longitudinal waves in the medium (but not in vacuum), and, in particular, the existence of electrical (without the magnetic component) longitudinal waves – see, for example, [19, page 73).

Let us note also that the existence of standing electromagnetic waves is also known. In [12] it is indicated that the standing electromagnetic waves are generated by vibrators and by some natural emitters.

10. Tesla Transformer as a fuel-free Energy Generator

There are numerous descriptions of experiments Tesla's power transmission and power generation. Nevertheless, one can find only one patent, relating to that subject, [22], and in him is no mention of Tesla transformer - they are described in other patents [23-27]. Thus, (as far is known the author) can not refer to a patent on the use Tesla transformer as the energy generator. However, some mention of his own successful experiments on the generation of energy by means of Tesla transformer [15].

It seems that the Tesla transformer as an energy generator at present is used most successfully in the inventions of Kapanadze [20], which are analyzed very thoroughly in [21]. Further we shall use the materials of [21] for considering the work of Kapanadze for the analysis from the position of the above theory.

Generator based on Tesla transformer may be presented in a simplified form shown on the Fig. 17.

DC current from an ordinary current source A is fed to the input of primary TT. Condenser C id discharged periodically by discharger P, and in the secondary coil of the primary TT an electric standing wave (as was shown above) is generated. The area of this wave W expands and reaches the secondary TT. The energy of this wave is transmitted to the secondary TT which is working in the mode of current generator. The generated current is rectified by rectifier M and is delivered to the load H (a part of the load power may be used instead of the current source A). It is important to note that the outputs of the secondary coils of both transformers are connected by a high-impedance wire R (as in the experiments of Avramenko). So the potentials of both coils are similar, and the currents in both coils are practically absent.

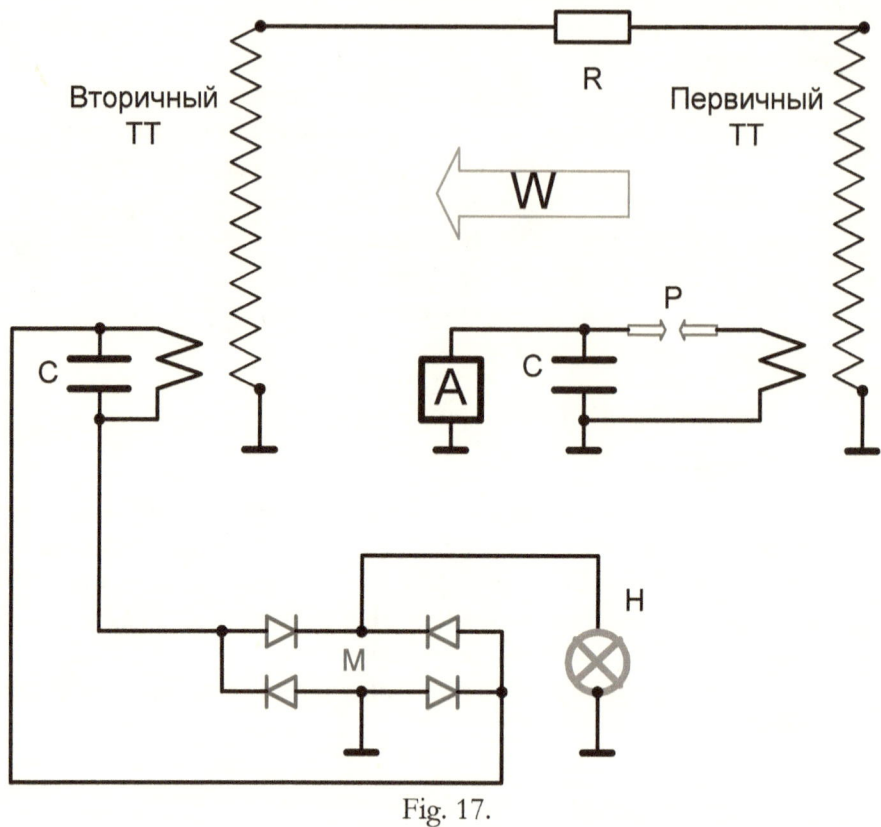

Fig. 17.

For the functioning of such device several conditions must be satisfied [21]:
1. the environment must supply energy,
2. the secondary transformer must generate current when the potential appears on its secondary coil,
3. the resonant frequency of both transformers should be similar.

The first condition in [21] was explained by the existence of ether. But above it was shown that the condition is fulfilled due to the fact that electric standing wave exists and is propagated with the aid of energy exchange with the air.

The second condition in [21] is explained by the reciprocity principle. We must note that in our case this principle "works" because not only the potential of primary transformer is being fed to the input of second TT, but also the standing wave that has kept all its characteristics in the secondary TT area. Indeed, if in the primary TT the problem in question is – to determine the electric potential and the intensities of electromagnetic wave with given primary current in the primary coil, in the secondary TT there is a problem of determining the current in the primary coil for given electric potential and intensity of an electromagnetic wave (and not only electrical potential). The intensities are called "given", as they "arrived" together with the standing wave. In Section 4 it was indicated that the Tesla transformer as a whole is described by equation system relating the variables

$$I_1, \ E_r(r,\varphi,y), \ E_y(r,\varphi,y), \ E_\varphi(r,\varphi,y), \ \varphi'(r,\varphi,y).$$

In the mode of wave generation the current I_1 is known, and the following functions are calculated

$$E_r(r,\varphi,y), \ E_y(r,\varphi,y), \ E_\varphi(r,\varphi,y), \ \varphi'(r,\varphi,y),$$

and in the mode of current I_1 generation these values are known.

The third condition is discussed in great detail in [21.] It is stated that the frequency and the phase of oscillations in the secondary TT can not be regulated, they can only be measured, but the frequency and the phase of oscillation in the primary TT should be adjusted in resonance with the measured values in the secondary TT. The complexity of technical implementation of this requirement is shown. Tesla strived to achieve the resonance (as shown in [21]) with the aid of complex mechanical controllers [28]. Kapanadze had found a solution using electronic circuits [20]. In [21] his solution is being analyzed. Evidently, it is exactly that solution which the other inventors do not see and cannot repeat .

Let us consider the energy balance in this circuit, using the following notations:

W_a - the energy of current source A,

W_c - the energy accumulated by the condenser before its discharge, накапливаемая конденсатором перед его разрядом,

W_1 - the energy expended in the primary transformer for the wave generation,

W_2 - the energy generated by the secondary transformer,

W_v - the energy of standing wave which is equal to the energy of thermal flow into the wave area,

W_H - the energy of load.

Presumably (as the author hadn't performed his own experiments) the energy variation may be presented (qualitatively) by the graphs depicted on Fig 18, where the graph of energy variation in the condenser is presented (first window), of the primary transformer (second window), electric wave (third window), and also the periods of the charge in condenser t_1 and of the charge in condenser t_2 are denoted. We have

$$\max W_1 = \max W_c,$$

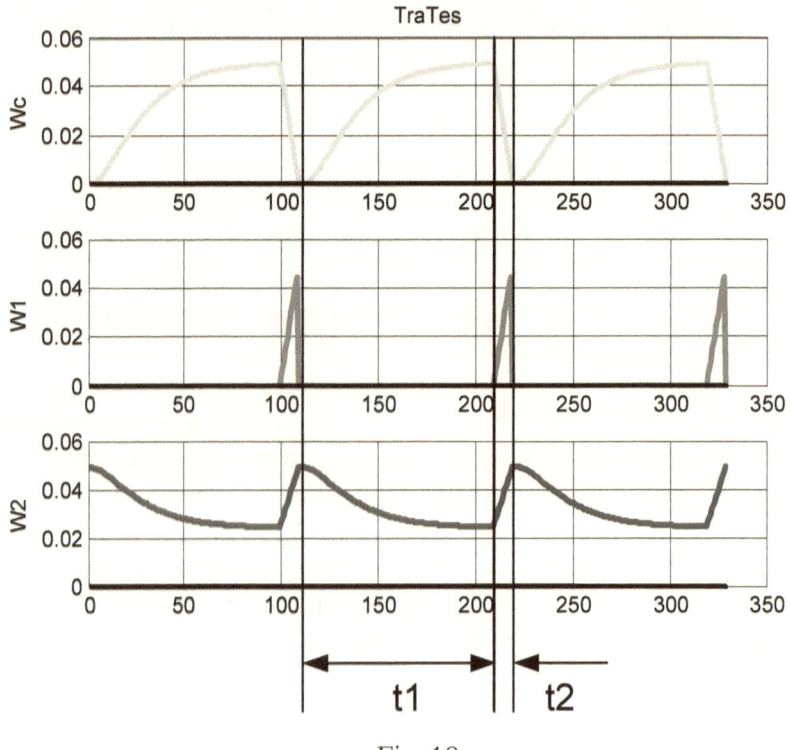

Fig. 18.

The wave's energy density in the period of the condenser t_1 charging is increasing with the wave's intensity increase, up to a certain value $\max W_v$, and then is decreasing in the period of condenser's t_2 discharge. to a certain value $\min W_v$ due to the instability of thermal processes in the wave area. The energy W_2 generated by the secondary transformer, and the equal to it energy of the load $W_H = W_2$, constitute some part of the wave's energy. We shall assume that $W_H = W_2 = k \cdot W_v$, где k is a certain coefficient.

The functions of currents (on a different time scale) of the primary coils of the transformer have a form shown on Fig. 19, where $i1$, $i2$ – are the currents of the primary and secondary transformers, accordingly.

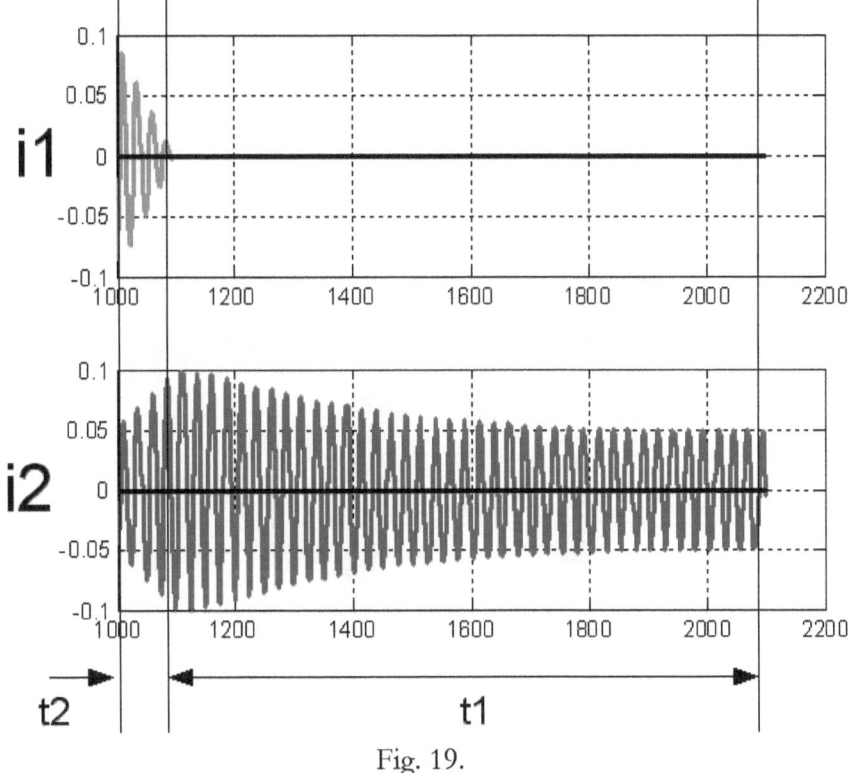

Fig. 19.

There are numerous known constructions of Don Smith – see, for instance [29, 30]. One of them is a Tesla transformer and several remote Tesla coils, identical to secondary coils of Tesla transformer see Fig. 20. According to the above said, we may assume that (as in the previous case)

- The primary coil of Tesla transformer is connected to a high frequency generator, and thus the Tesla transformer generates a standing electric wave,
- The energy of this wave (replenished by the energy of thermal flow from the environment) is transmitted to the remote coils – their number may be arbitrary,

- each remote coil is connected to the condenser, and all the circuit is connected to the load through high-voltage diodes; in this way each of such circuits operates as a generator of current.

As in the previous case, the circuit of the remote coil must be set in resonance with the frequency of Tesla transformer, i.e., with the wave's frequency. However, in Don Smith's works we do not see any complicated circuits for resonance setting. Evidently, in the device of Kapanadze the main frequency destabilizing factor may have been the mutual inductivity of the secondary Tesla transformer, which is determined by the characteristics of the non-stable air interval between the coils. In Smith's construction this mutual inductivity in the remote coil circuit is absent.

Fig. 20.

11. Energy Transfer by Tesla Flat Coils

[31, 32] describes the technology of energy transmission on a distance, which was called by the authors "wireless electricity" (WiTricity). This technology make use of energy transmission between Tesla flat coils, which is evident from the system structure – see Fig. 21, where one of the pilot units. The authors note the following features of the technology:
- low efficiency of the transmission (from 40% to 95%),
- increase in efficiency at increase in load,
- transfered power – up to 3 kWt,
- transmission range – several meters.
- high frequency of transmissing magnetic field,

- low intensity of the magnetic field (magnetic induction $\approx 3 \cdot 10^{-5} T$),
- the related safety of the user,
- ratio of the transmitting coil's area to the receiving coil's area is about 15.

Fig. 21.

The authors do not find any explanation to some features of their system, and, in particular, to the efficiency increase with the load increase. However, all these features are easily explained, if we take into account the above. Especially, the reason for the efficiency increase on load increase is that the receiving coils also form a standing wave and thereby are catalyzing the increase of thermal flow from the environment.

The authors do not show the current strength in the coil, but it is clear that the magnetic field's induction can be increased by two orders. The efficiency can exceed 100%, which means that the energy can be extracted from the environment. Such experiments are to be found in Internet – see, for instant, [34] and Fig. 22.

The described technology – "Historicity" is using flat coils, for which the intensity of standing magnetic wave does not depend on the distance to the coil's plane. That's why the cylindrical coils for which the intensity of standing magnetic wave decays hyperbolically depending on the distance to the cylinder surface, cannot be used in this technology.

However, two coaxial cylindrical coils with small gap between them can transfer the energy to each other. The gap should be aerial. It seems that such structure will be able to extract energy from the environment.

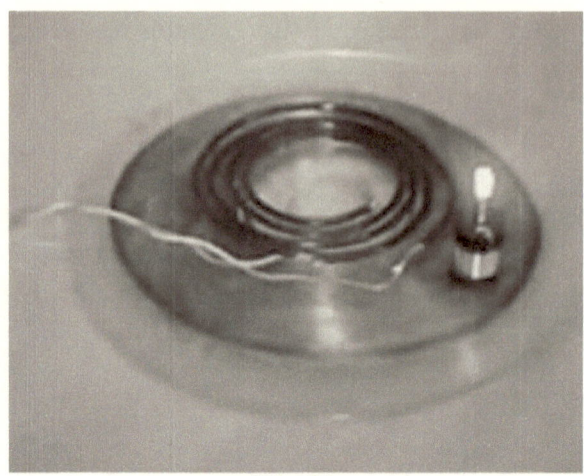

Fig. 22.

References

1. Zaew N.E. Superconductors engineer Avramenko. Journal "Technika - molodegi", №1, Moscow., 1991, in Russian,
2. Kuligin V.A., Kornewa M.V., Kuligina G.A., Bolshakow G.P. Non-inertial charges and currents, in Russian, http://www.n-t.ru/ac/iga/
3. M. Lobova, G. Shipov, Tawatchai Laosirihongthong, Supakit Chotigo. Experimental observation of the scalar electromagnetic field. King Mongkut's University of Technology, Thonbury, Bangkok, 10140, Thailand. (in Internet, file 1032-MonopolE.pdf)
4. Zaew N.E., Awramenko S.V., Lisin V.N. Measurement of the conduction current, excited by the polarization current, in Russian, http://rusphysics.ru/dissertation/269/
5. Forum "A revolution in the power sector: Russian engineers have repeated the installation of Tesla?" Number 24, PVA (June 27, 2011 9:35), in Russian, http://oko-planet.su/phenomen/phenomenscience/page,1,1,72768-revolyuciya-v-elektroenergetike-rossiyskie-inzhenery-sozdali.html#comment
6. Khmelnik S.I. Calculation of static electric and magnetic fields on the basis of a variational principle. "Papers of Independent

Authors", publ. «DNA», printed in USA, Lulu Inc. 11744286. Israel-Russia, 2011, iss. 19, ISBN 978-1-105-15373-0 (in Russian).
7. Гольдштейн Л.Д., Зернов Н.В. Электромагнитные поля и волны. Издание второе, переработанное и дополненное. Изд. "Советское радио", Москва, 1971. – 665 с.
8. Khmelnik S.I. Variational Principle of Extremum in Electromechanical and Electrodynamic Systems. Publisher by "MiC", printed in USA, Lulu Inc., ID 1142842, Israel, 2008, ISBN 978-0-557-08231-5.
9. Khmelnik S.I. Energy processes in free-full electromagnetic generators, second edition, Publisher by "MiC", printed in USA, Lulu Inc., ID 10292524, Израиль, 2011, ISBN 978-1-257-05555-5.
10. Roshchin V.V., Godin S. M. Experimental Investigation of Physical Effects in Dynamic Magnetic System. Letters to Journal of Theoretical Physics, 2000, volume 26, iss. 24 (in Russian),
http://www.ioffe.rssi.ru/journals/pjtf/2000/24/p70-75.pdf
11. B.M. Jaworski, A.A. Detlaph. Справочник по физике, Moscow, Publ. "Fizmatlit", 1963, in Russian.
12. Verin O.G. The theory of Tesla transformer, in Russian.
http://www.sciteclibrary.ru/rus/catalog/pages/10404.html
13. Peter Lindemann. Free energy, http://www.free-energy.ws/peter-lindemann.html
14. Kosinov N.V. Experiments on the wireless transmission of energy: the confirmation of the revolutionary ideas of N. Tesla, in Russian, http://kosinov.314159.ru/kosinov31.htm
15. Katargin R.K. Legacy of Nikola Tesla, in Russian.
http://forum.lah.ru/_fr/21/Tesla-Kap.pdf
16. A.A. Detlaph, B.M. Jaworski, L.B. Milkovskaya. Physics Course, Volume 1, Electricity and Magnetism, fourth edition, Moscow, Publ. "High School", 1977, in Russian.
17. Jaworski B.M., Pinsky A.A. Foundations of Physics. V.1. Mech. Molecular Physics. Electrodynamics. Moscow, Publ. "Fizmatlit", 2003, in Russian.
18. Isaev S.I., Kozhinov I.A. etc. The theory of heat exchange. Moscow, publ. «Vyshya shkola», 1979, 495 p., in Russian.
19. Vinogradova M.B., Rudenko O.V., Suhorukov A.P. Theory of wave. Moscow, publ. "Nauka", 1979, in Russian.
20. Fuel-less generator Kapanadze, Internet.

21. Zarew V.A. Installing Tariel Kapanadze (reconstruction), in Russian, http://halerman.narod.ru/TTCG/Kapanadze.htm
22. Nicola Tesla. Art of transmitting electrical energy through the natural mediums. USPO, 1905, Patent 787,412
23. Nicola Tesla. Method of regulating apparatus for producing currents of high frequency. USPO, 1896, Patent 568,178.
24. Nicola Tesla. Method of and apparatus for producing currents of high frequency. USPO, 1896, Patent 568,179.
25. Nicola Tesla. Apparatus for producing electrical currents of high frequency. USPO, 1896, Patent 568,180.
26. Nicola Tesla. Apparatus for producing electric currents of high frequency. USPO, 1897, Patent 577,670.
27. Nicola Tesla. Apparatus for producing currents of high frequency. USPO, 1897, Patent 583,953.
28. Nicola Tesla. Electric-circuit Controller. USPO, 1898, Patents 613,735; 611,719; 609,251; 609,248; 609,247; 609,249; 609,245; 609,246.
29. Donald L. Smith. Resonanse Energy Method, 2002, http://www.free-energy-info.co.uk/Smith.pdf
30. Practical Guide to Free-Energy Devices, http://www.free-energy-info.co.uk/
31. Wireless electricity is struck by its creators, http://www.membrana.ru/particle/1986
32. André Kurs, Robert Moffatt, and Marin Soljačić. Simultaneous mid-range power transfer to multiple devices. Appl. Phys. Lett. 96, 044102 (2010), http://apl.aip.org/resource/1/applab/v96/i4/p044102_s1?isAuthorized=no
33. The scheme of the simplest transformer Tesla, http://shemalog.narod.ru/pn2.html
34. The flat Tesla coil, http://www.youtube.com/watch?v=444j9N3G--U

www.ingramcontent.com/pod-product-compliance
Lightning Source LLC
Chambersburg PA
CBHW021929170526
45157CB00005B/2247